My Grandma Told Me

By Kelley E Howe Nolan

Illustrations by Stephanie Wenning and Kelley E Howe Nolan

Library of Congress Control Number 18737114911.

Cover Design/Formatting by Anna Nelson, One Small Step Writing Services.

ISBN 978-1-7369311-0-3 (paperback)
ISBN 978-1-7369311-1-0 (ebook)

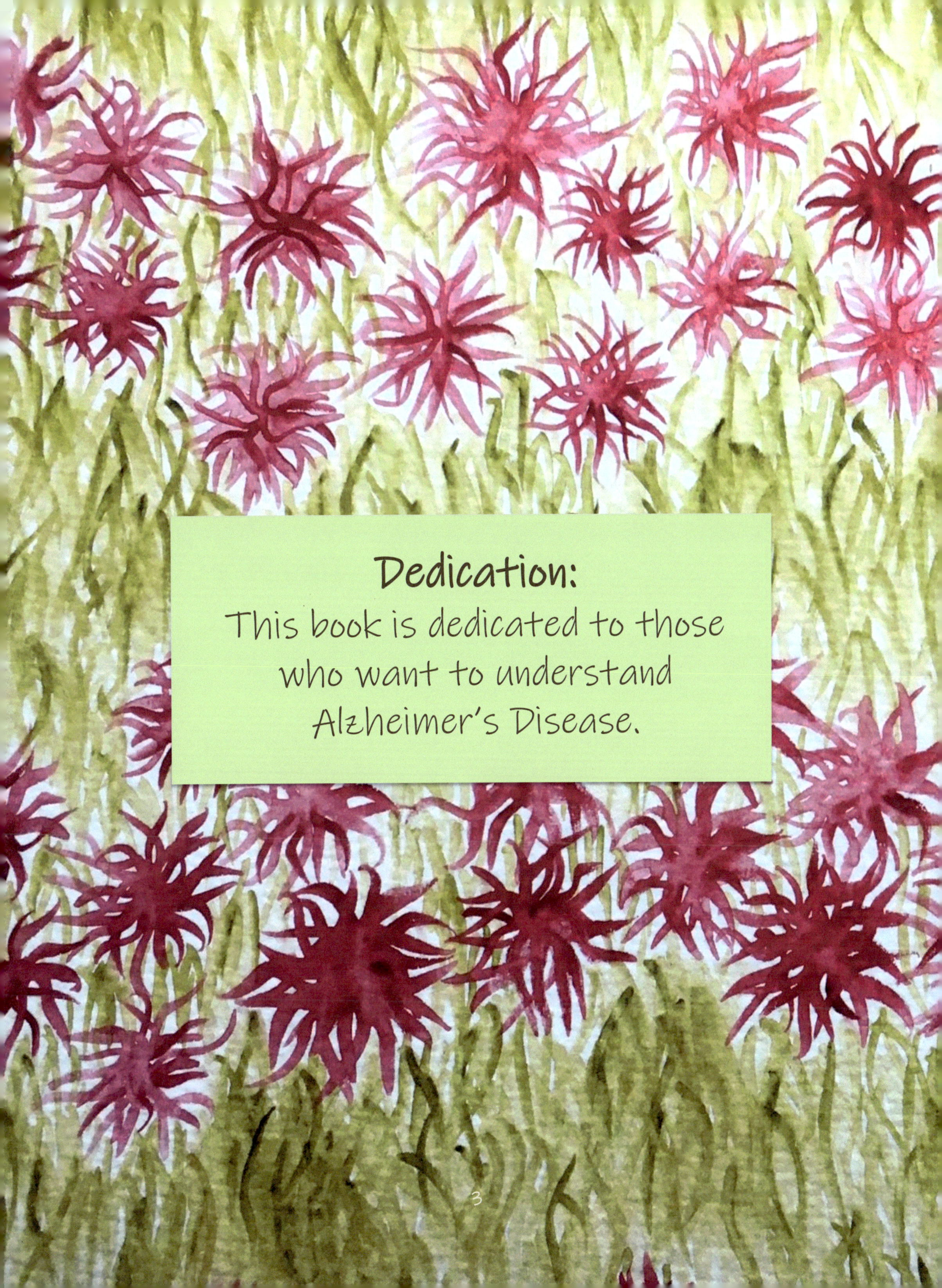

Dedication:

This book is dedicated to those who want to understand Alzheimer's Disease.

One day Kamryn and her grandma were outside in grandma's front yard. They were looking at all the beautiful flowers.

Grandma asked Kamryn if she knew that grandpa has Alzheimer's Disease?

Kamryn told grandma that she didn't know what that was. What is it?

Grandma asked Kamryn if she remembers how grandpa sometimes seems to forget things?

Kamryn told grandma she remembers. Like when grandpa can't find his glasses or where he put his soda?

Grandma told Kamryn that it's not because grandpa just forgets. His brain is changing.

Kamryn asked grandma,
"what do you mean - what's
happening to grandpa's brain?"

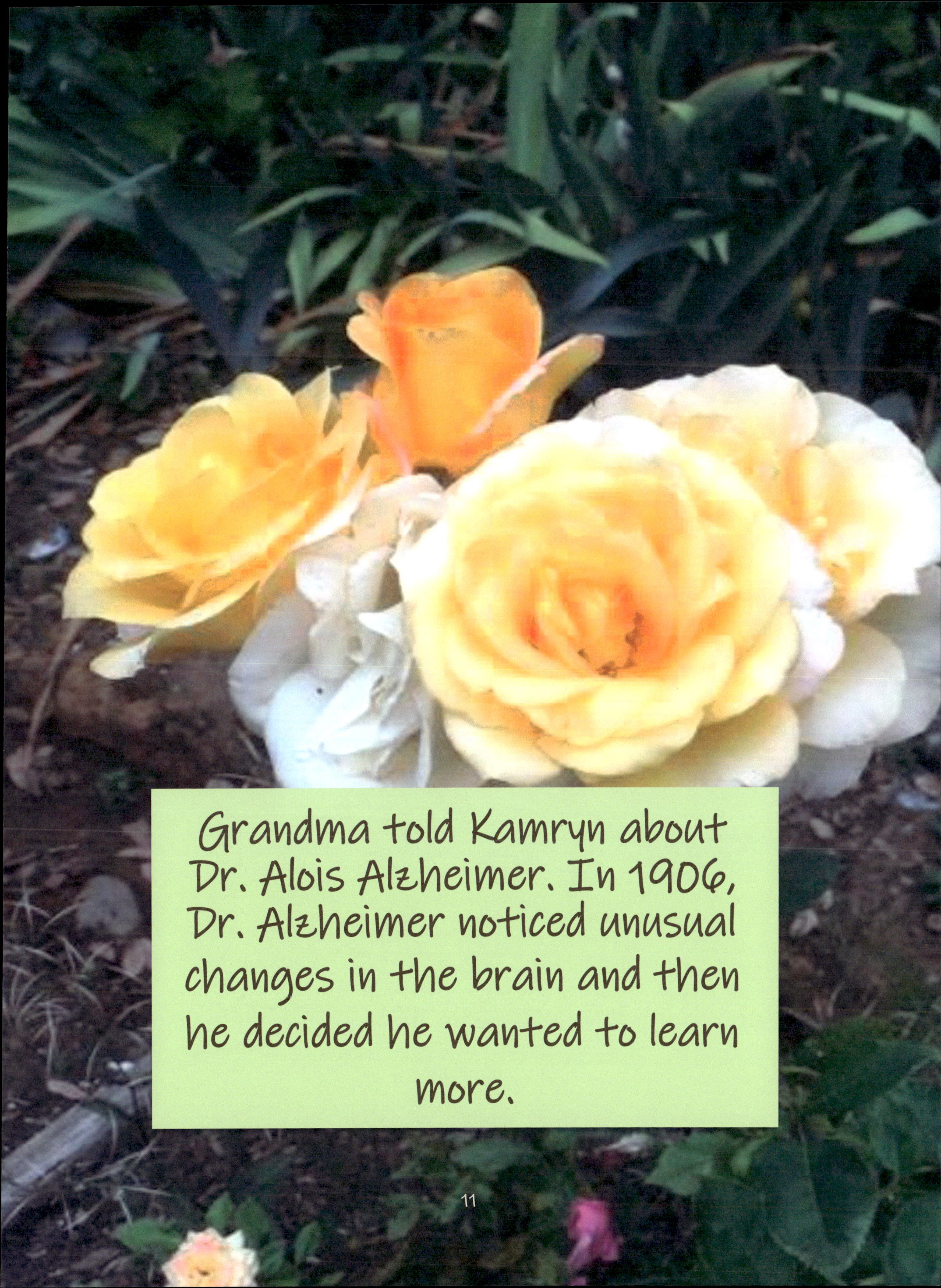

Grandma told Kamryn about Dr. Alois Alzheimer. In 1906, Dr. Alzheimer noticed unusual changes in the brain and then he decided he wanted to learn more.

Grandma told Kamryn that this disease is in grandpa's brain and it changes his memory.

Kamryn looked at her grandma and she said, "what are we going to do about it?"

Grandma hugged her granddaughter and said, "We're going to learn about science. We're going to start learning about the brain so we can understand Alzheimer's Disease."

So, grandma and Kamryn began to look at all the different parts of the brain.

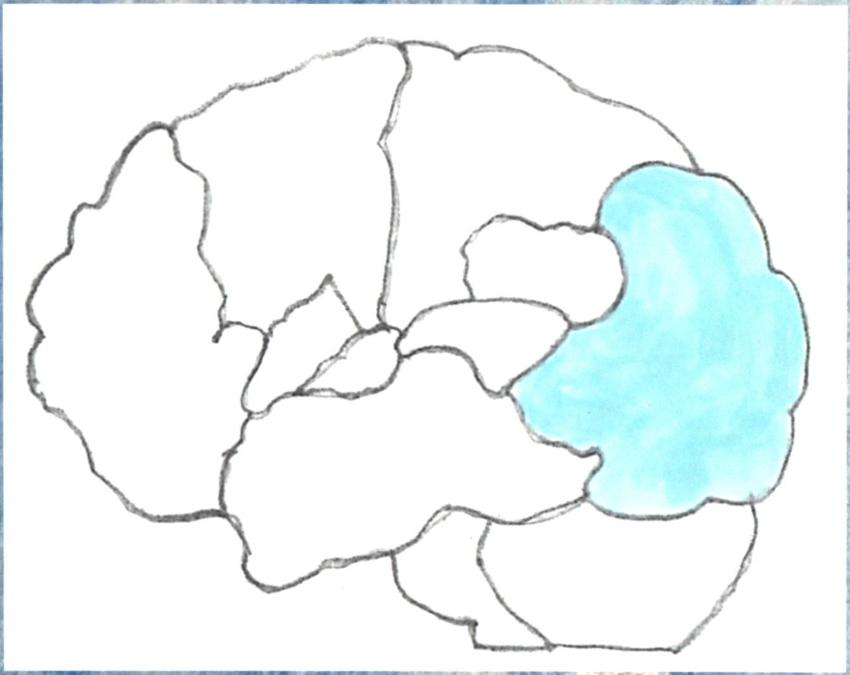

They began with the area of vision.

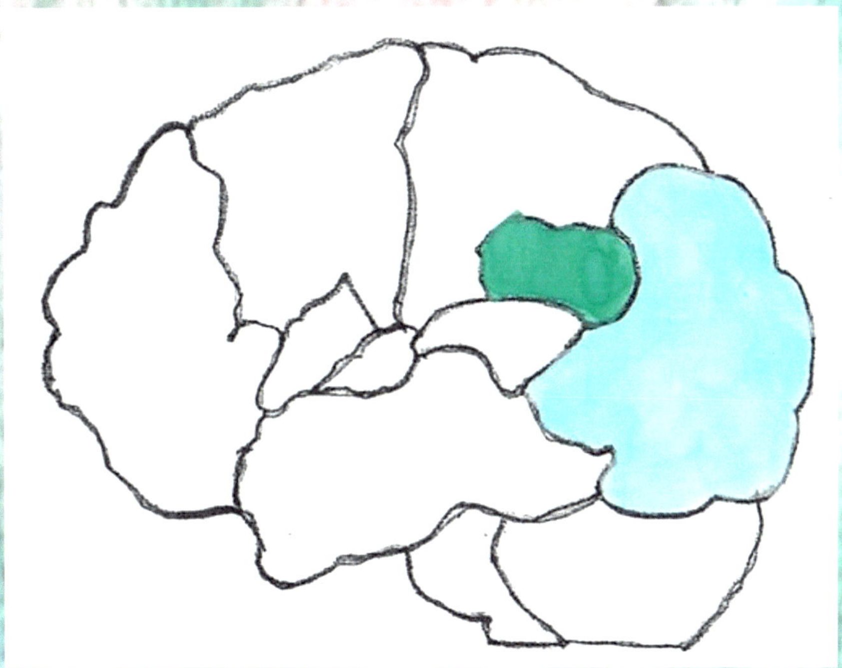

Taste.

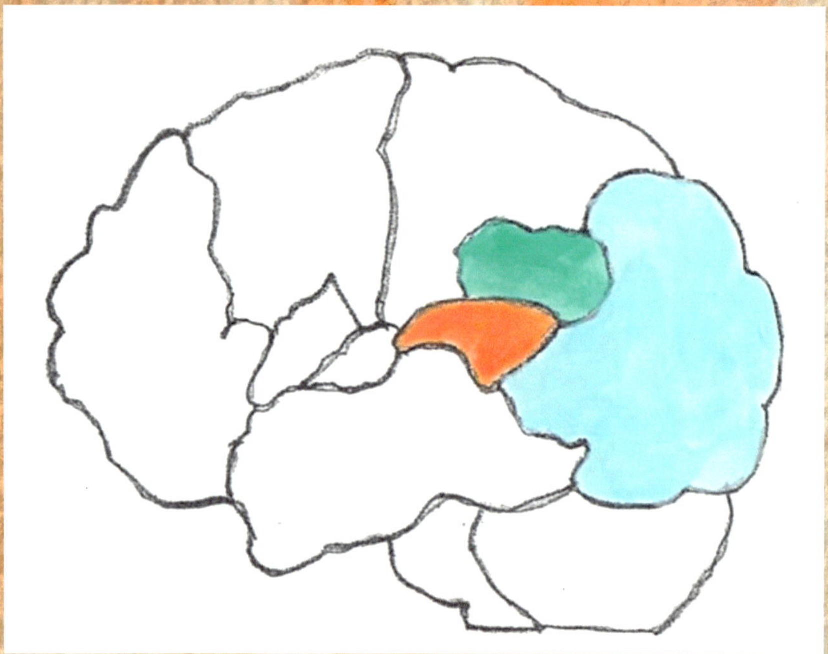

Hearing.

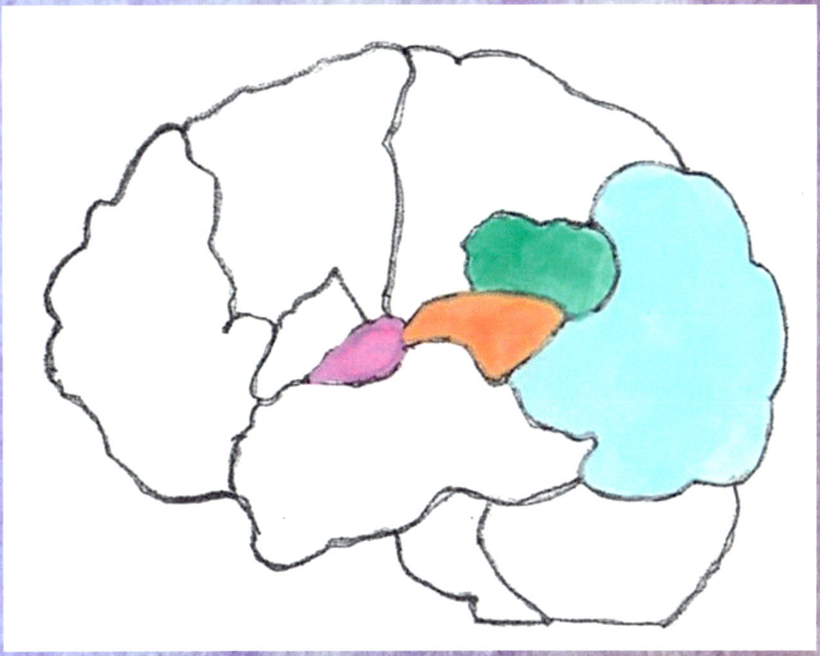

Smell.

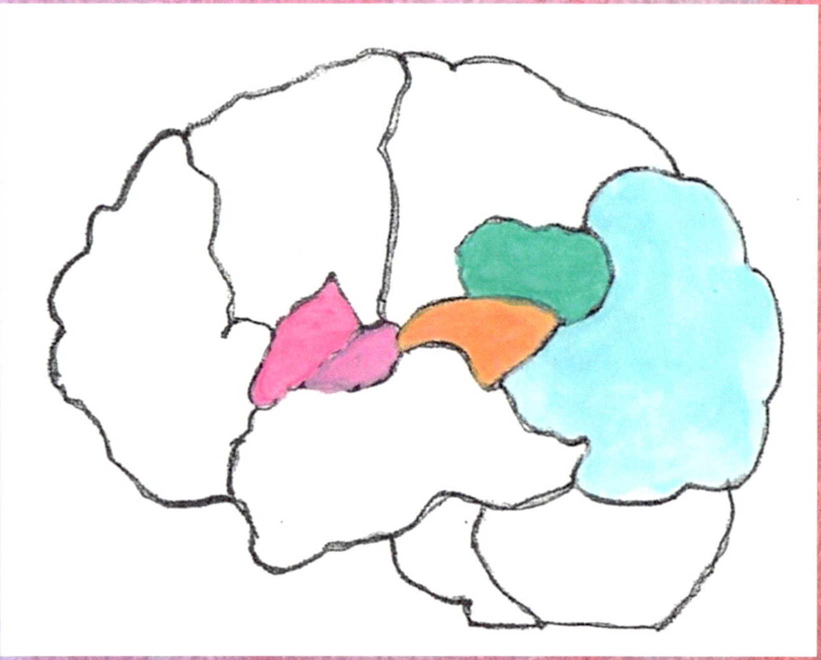

Speech.

Touch.

Movement.

Decision Making; Planning; Attention.

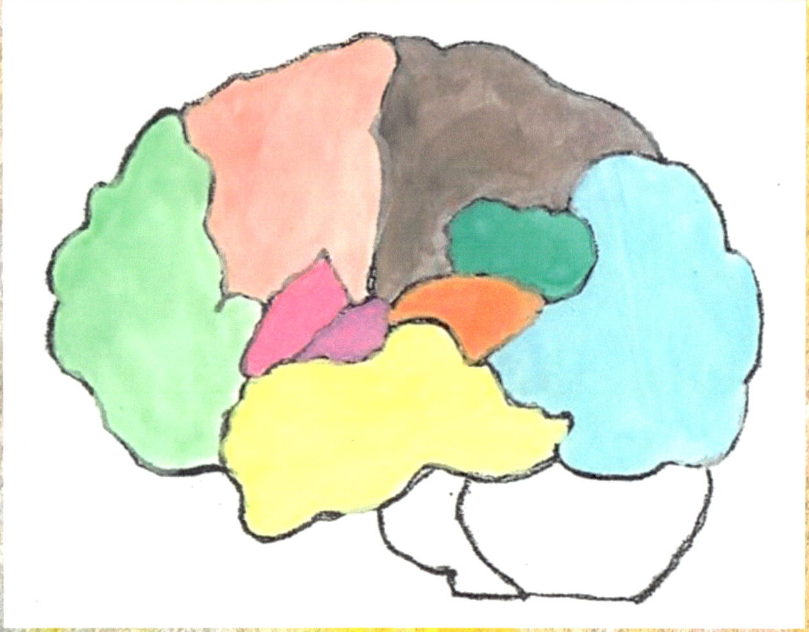

Emotion; Memory.

Kamryn looked at her grandma
and asked, "Grandma,
will grandpa forget you?"

Grandma held Kamryn and said, "Yes, grandpa might forget who we are in his brain, but Kamryn, in his heart he will always know who we are, and he will always love us.

The following is a list of children's books that you may find as additional reading on Alzheimer's Disease and other memory loss:

1. Abeele, Veronique Van den. Still My Grandma. Grand Rapids: First published by Mijade in Belgium then Wm. B. Eerdmans Publishing Co., 2006.
2. Altman, Linda Jacobs. Singing with Momma Lou. New York: Lee & Low Books Inc., 2002.
3. Bahr, Mary. The Memory Box. Chicago: Albert Whitman & Company, 1992.
4. Contreras, Kathleen. Sweet Memories/Dulces Recuerdos. Lyndhurst: Lectorum Publications, Inc., 2014.
5. Oliveros, Jessie. The Remember Balloons. New York: Simon & Schuster Children's Publishing Division, 2018.
6. Scaccco, Linda. Always My Grandpa - A Story for Children About Alzheimer's Disease. Washington : Magination Press, 2006.
7. Schnurbush, Barbara. Striped Shirts and Flowered Pants. Washington: Magination Press, 2007.
8. Shriver, Maria. What's Happening to Grandpa? New York: Little, Brown and Company and Warner Books, 2004.

The following books are focused on discussing the brain with children of various ages:

1. BabyProfessor. My Little Brain! Explaining The Human Brain for Kids. 1 ed., Newark, DE, Speedy Publishing LLC, 2016.
2. Deak, PhD, JoAnn. Your FANTASTIC ELASTIC BRAIN Stretch It, Shape It. Naperville, Illinois, Little Pickle Press, an imprint of Sourcebooks Jabberwocky, 2010.
3. Matt, Margaret, and Joe Ziemian. Human Anatomy Coloring Book. First ed., New York, Dover Publications, Inc., 1982.
4. Seluk, Nick. The Brain is Kind of a Big Deal. First Edition ed., vol. 1, New York, New York, Orchard Books / Scholastic, October 2019. 1 vols.
5. Silver, Donald M., and Patricia J. Wynne. My First Book About the BRAIN. Mineola, New Yort, Dover Publications Inc., 2013.

Acknowledgements:

I'd like to thank my family and friends for their love and support during this time in my life. I am especially thankful for the friendship of the late Dr. Albert Halls, who gave me his time and knowledge about the brain, without which I would not have been able to learn the impact it plays on this disease.

Made in the USA
Las Vegas, NV
07 November 2026

57650026R00019